Learn Data Analysis with Python

Lessons in Coding

A.J. Henley
Dave Wolf

Apress®

Learn Data Analysis with Python: Lessons in Coding

A.J. Henley
Washington, D.C.,
District of Columbia,
USA

Dave Wolf
Adamstown,
Maryland,
USA

ISBN-13 (pbk): 978-1-4842-3485-3
https://doi.org/10.1007/978-1-4842-3486-0

ISBN-13 (electronic): 978-1-4842-3486-0

Library of Congress Control Number: 2018933537

Copyright © 2018 by A.J. Henley and Dave Wolf

This work is subject to copyright. All rights are reserved by the Publisher, whether the whole or part of the material is concerned, specifically the rights of translation, reprinting, reuse of illustrations, recitation, broadcasting, reproduction on microfilms or in any other physical way, and transmission or information storage and retrieval, electronic adaptation, computer software, or by similar or dissimilar methodology now known or hereafter developed.

Trademarked names, logos, and images may appear in this book. Rather than use a trademark symbol with every occurrence of a trademarked name, logo, or image we use the names, logos, and images only in an editorial fashion and to the benefit of the trademark owner, with no intention of infringement of the trademark.

The use in this publication of trade names, trademarks, service marks, and similar terms, even if they are not identified as such, is not to be taken as an expression of opinion as to whether or not they are subject to proprietary rights.

While the advice and information in this book are believed to be true and accurate at the date of publication, neither the authors nor the editors nor the publisher can accept any legal responsibility for any errors or omissions that may be made. The publisher makes no warranty, express or implied, with respect to the material contained herein.

Managing Director, Apress Media LLC: Welmoed Spahr
Acquisitions Editor: Steve Anglin
Development Editor: Matthew Moodie
Coordinating Editor: Mark Powers

Cover designed by eStudioCalamar

Cover image designed by Freepik (www.freepik.com)

Distributed to the book trade worldwide by Springer Science+Business Media New York, 233 Spring Street, 6th Floor, New York, NY 10013. Phone 1-800-SPRINGER, fax (201) 348-4505, email orders-ny@springer-sbm.com, or visit www.springeronline.com. Apress Media, LLC is a California LLC and the sole member (owner) is Springer Science + Business Media Finance Inc (SSBM Finance Inc). SSBM Finance Inc is a **Delaware** corporation.

For information on translations, please email editorial@apress.com; for reprint, paperback, or audio rights, please email bookpermissions@springernature.com.

Apress titles may be purchased in bulk for academic, corporate, or promotional use. eBook versions and licenses are also available for most titles. For more information, reference our Print and eBook Bulk Sales web page at http://www.apress.com/bulk-sales.

Any source code or other supplementary material referenced by the author in this book is available to readers on GitHub via the book's product page, located at www.apress.com/9781484234853. For more detailed information, please visit http://www.apress.com/source-code.

Printed on acid-free paper

Table of Contents

About the Authors

A.J. Henley is a technology educator with over 20 years' experience as a developer, designer, and systems engineer. He is an instructor at both Howard University and Montgomery College.

Dave Wolf is a certified Project Management Professional (PMP) with over 20 years' experience as a software developer, analyst, and trainer. His latest projects include collaboratively developing training materials and programming bootcamps for Java and Python.

About the Technical Reviewer

 **Michael Thomas** has worked in software
development for more than 20 years as an
individual contributor, team lead, program
manager, and vice president of engineering.
Michael has more than ten years of experience
working with mobile devices. His current focus
is in the medical sector, using mobile devices
to accelerate information transfer between
patients and health-care providers.

## CHAPTER 1

How to Use This Book

If you are already using Python for data analysis, just browse this book's table of contents. You will probably find a bunch of things that you wish you knew how to do in Python. If so, feel free to turn directly to that chapter and get to work. Each lesson is, as much as possible, self-contained.

Be warned! This book is more a workbook than a textbook.

If you aren't using Python for data analysis, begin at the beginning. If you work your way through the whole workbook, you should have a better of idea of how to use Python for data analysis when you are done.

If you know nothing at all about data analysis, this workbook might not be the place to start. However, give it a try and see how it works for you.

Installing Jupyter Notebook

The fastest way to install and use Python is to do what you already know how to do, and you know how to use your browser. Why not use Jupyter Notebook?

© A.J. Henley and Dave Wolf 2018
A.J. Henley and D. Wolf, *Learn Data Analysis with Python*,
https://doi.org/10.1007/978-1-4842-3486-0_1

What Is Jupyter Notebook?

Jupyter Notebook is an interactive Python shell that runs in your browser. When installed through Anaconda, it is easy to quickly set up a Python development environment. Since it's easy to set up and easy to run, it will be easy to learn Python.

Jupyter Notebook turns your browser into a Python development environment. The only thing you have to install is Anaconda. In essence, it allows you to enter a few lines of Python code, press CTRL+Enter, and execute the code. You enter the code in cells and then run the currently selected cell. There are also options to run all the cells in your notebook. This is useful if you are developing a larger program.

What Is Anaconda?

Anaconda is the easiest way to ensure that you don't spend all day installing Jupyter. Simply download the Anaconda package and run the installer. The Anaconda software package contains everything you need to create a Python development environment. Anaconda comes in two versions—one for Python 2.7 and one for Python 3.x. For the purposes of this guide, install the one for Python 2.7.

Anaconda is an open source data-science platform. It contains over 100 packages for use with Python, R, and Scala. You can download and install Anaconda quickly with minimal effort. Once installed, you can update the packages or Python version or create environments for different projects.

Getting Started

1. Download and install Anaconda at `https://www.anaconda.com/download`.

2. Once you've installed Anaconda, you're ready to create your first notebook. Run the Jupyter Notebook application that was installed as part of Anaconda.

3. Your browser will open to the following address: `http://localhost:8888`. If you're running Internet Explorer, close it. Use Firefox or Chrome for best results. From there, browse to `http://localhost:8888`.

4. Start a new notebook. On the right-hand side of the browser, click the drop-down button that says "New" and select *Python* or *Python 2*.

5. This will open a new iPython notebook in another browser tab. You can have many notebooks open in many tabs.

6. Jupyter Notebook contains cells. You can type Python code in each cell. To get started (for Python 2.7), type `print "Hello, World!"` in the first cell and hit CTRL+Enter. If you're using Python 3.5, then the command is `print("Hello, World!")`.

Getting the Datasets for the Workbook's Exercises

1. Download the dataset files from http://www.ajhenley.com/dwnld.

2. Upload the file datasets.zip to Anaconda in the same folder as your notebook.

3. Run the Python code in Listing 1-1 to unzip the datasets.

Listing 1-1. Unzipping dataset.zip

```python
path_to_zip_file = "datasets.zip"
directory_to_extract_to = ""

import zipfile
zip_ref = zipfile.ZipFile(path_to_zip_file, 'r')
zip_ref.extractall(directory_to_extract_to)
zip_ref.close()
```

CHAPTER 2

Getting Data Into and Out of Python

The first stage of data analysis is getting the data. Moving your data from where you have it stored into your analytical tools and back out again can be a difficult task if you don't know what you are doing. Python and its libraries try to make it as easy as possible.

With just a few lines of code, you will be able to import and export data in the following formats:

- CSV
- Excel
- SQL

Loading Data from CSV Files

Normally, data will come to us as files or database links. See Listing 2-1 to learn how to load data from a CSV file.

Listing 2-1. Loading Data from CSV File

```
import pandas as pd
Location = "datasets/smallgradesh.csv"
df = pd.read_csv(Location, header=None)
```

© A.J. Henley and Dave Wolf 2018
A.J. Henley and D. Wolf, *Learn Data Analysis with Python*,
https://doi.org/10.1007/978-1-4842-3486-0_2

Now, let's take a look at what our data looks like (Listing 2-2):

Listing 2-2. Display First Five Lines of Data

```
df.head()
```

As you can see, our dataframe lacks column headers. Or, rather, there are headers, but they weren't loaded as headers; they were loaded as row one of your data. To load data that includes headers, you can use the code shown in Listing 2-3.

Listing 2-3. Loading Data from CSV File with Headers

```
import pandas as pd
Location = "datasets/gradedata.csv"
df = pd.read_csv(Location)
```

Then, as before, we take a look at what the data looks like by running the code shown in Listing 2-4.

Listing 2-4. Display First Five Lines of Data

```
df.head()
```

If you have a dataset that doesn't include headers, you can add them afterward. To add them, we can use one of the options shown in Listing 2-5.

Listing 2-5. Loading Data from CSV File and Adding Headers

```
import pandas as pd
Location = "datasets/smallgrades.csv"
# To add headers as we load the data...
df = pd.read_csv(Location, names=['Names','Grades'])
# To add headers to a dataframe
df.columns = ['Names','Grades']
```

Your Turn

Can you make a dataframe from a file you have uploaded and imported on your own? Let's find out. Go to the following website, which contains U.S. Census data (http://census.ire.org/data/bulkdata.html), and download the CSV datafile for a state. Now, try to import that data into Python.

Saving Data to CSV

Maybe you want to save your progress when analyzing data. Maybe you are just using Python to massage some data for later analysis in another tool. Or maybe you have some other reason to export your dataframe to a CSV file. The code shown in Listing 2-6 is an example of how to do this.

Listing 2-6. Exporting a Dataset to CSV

```
import pandas as pd
names = ['Bob','Jessica','Mary','John','Mel']
grades = [76,95,77,78,99]
GradeList = zip(names,grades)
df = pd.DataFrame(data = GradeList, columns=['Names','Grades'])

df.to_csv('studentgrades.csv',index=False,header=False)
```

Lines 1 to 6 are the lines that create the dataframe. Line 7 is the code to export the dataframe df to a CSV file called studentgrades.csv.

The only parameters we use are index and header. Setting these parameters to false will prevent the index and header names from being exported. Change the values of these parameters to get a better understanding of their use.

If you want in-depth information about the to_csv method, you can, of course, use the code shown in Listing 2-7.

Listing 2-7. Getting Help on to_csv

```
df.to_csv?
```

Your Turn

Can you export the dataframe created by the code in Listing 2-8 to CSV?

Listing 2-8. Creating a Dataset for the Exercise

```
import pandas as pd
names = ['Bob','Jessica','Mary','John','Mel']
grades = [76,95,77,78,99]
bsdegrees = [1,1,0,0,1]
msdegrees = [2,1,0,0,0]
phddegrees = [0,1,0,0,0]
Degrees = zip(names,grades,bsdegrees,msdegrees,phddegrees)
columns = ['Names','Grades','BS','MS','PhD']
df = pd.DataFrame(data = Degrees, columns=column)
df
```

Loading Data from Excel Files

Normally, data will come to us as files or database links. Let's see how to load data from an Excel file (Listing 2-9).

Listing 2-9. Loading Data from Excel File

```
import pandas as pd
Location = "datasets/gradedata.xlsx"
df = pd.read_excel(Location)
```

Now, let's take a look at what our data looks like (Listing 2-10).

Listing 2-10. Display First Five Lines of Data

```
df.head()
```

If you wish to change or simplify your column names, you can run the code shown in Listing 2-11.

Listing 2-11. Changing Column Names

```
df.columns = ['first','last','sex','age','exer','hrs','grd','addr']
df.head()
```

Your Turn

Can you make a dataframe from a file you have uploaded and imported on your own? Let's find out. Go to https://www.census.gov/support/USACdataDownloads.html and download one of the Excel datafiles at the bottom of the page. Now, try to import that data into Python.

Saving Data to Excel Files

The code shown in Listing 2-12 is an example of how to do this.

Listing 2-12. Exporting a Dataframe to Excel

```python
import pandas as pd
names = ['Bob','Jessica','Mary','John','Mel']
grades = [76,95,77,78,99]
GradeList = zip(names,grades)
df = pd.DataFrame(data = GradeList,
        columns=['Names','Grades'])
writer = pd.ExcelWriter('dataframe.xlsx', engine='xlsxwriter')
df.to_excel(writer, sheet_name='Sheet1')
writer.save()
```

If you wish, you can save different dataframes to different sheets, and with one .save() you will create an Excel file with multiple worksheets (see Listing 2-13).

Listing 2-13. Exporting Multiple Dataframes to Excel

```python
writer = pd.ExcelWriter('dataframe.xlsx',engine='xlsxwriter')
df.to_excel(writer, sheet_name='Sheet1')
df2.to_excel(writer, sheet_name='Sheet2')
writer.save()
```

Note   This assumes that you have another dataframe already loaded into the df2 variable.

Your Turn

Can you export the dataframe created by the code shown in Listing 2-14 to Excel?

Listing 2-14. Creating a Dataset for the Exercise

```
import pandas as pd
names = ['Nike','Adidas','New Balance','Puma','Reebok']
grades = [176,59,47,38,99]
PriceList = zip(names,prices)
df = pd.DataFrame(data = PriceList, columns=['Names','Prices'])
```

Combining Data from Multiple Excel Files

In earlier lessons, we opened single files and put their data into individual dataframes. Sometimes we will need to combine the data from several Excel files into the same dataframe.

We can do this either the long way or the short way. First, let's see the long way (Listing 2-15).

Listing 2-15. Long Way

```
import pandas as pd
import numpy as np

all_data = pd.DataFrame()

df = pd.read_excel("datasets/data1.xlsx")
all_data = all_data.append(df,ignore_index=True)

df = pd.read_excel("datasets/data2.xlsx")
all_data = all_data.append(df,ignore_index=True)
```

```
df = pd.read_excel("datasets/data3.xlsx")
all_data = all_data.append(df,ignore_index=True)
all_data.describe()
```

- Line 4: First, let's set all_data to an empty dataframe.

- Line 6: Load the first Excel file into the dataframe df.

- Line 7: Append the contents of df to the dataframe all_data.

- Lines 9 & 10: Basically the same as lines 6 & 7, but for the next Excel file.

Why do we call this the long way? Because if we were loading a hundred files instead of three, it would take hundreds of lines of code to do it this way. In the words of my friends in the startup community, it doesn't scale well. The short way, however, does scale.

Now, let's see the short way (Listing 2-16).

Listing 2-16. Short Way

```
import pandas as pd
import numpy as np
import glob

all_data = pd.DataFrame()
for f in glob.glob("datasets/data*.xlsx"):
    df = pd.read_excel(f)
    all_data = all_data.append(df,ignore_index=True)
all_data.describe()
```

- Line 3: Import the glob library.

- Line 5: Let's set all_data to an empty dataframe.

- Line 6: This line will loop through all files that match the pattern.

- Line 7: Load the Excel file in f into the dataframe df.

- Line 8: Append the contents of df to the dataframe all_data.

Since we only have three datafiles, the difference in code isn't that noticeable. However, if we were loading a hundred files, the difference in the amount of code would be huge. This code will load all the Excel files whose names begin with data that are in the datasets directory no matter how many there are.

Your Turn

In the datasets/weekly_call_data folder, there are 104 files of weekly call data for two years. Your task is to try to load all of that data into one dataframe.

Loading Data from SQL

Normally, our data will come to us as files or database links. Let's learn how to load our data from a sqlite database file (Listing 2-17).

Listing 2-17. Load Data from sqlite

```
import pandas as pd
from sqlalchemy import create_engine
# Connect to sqlite db
db_file = r'datasets/gradedata.db'
engine = create_engine(r"sqlite:///{}"
        .format(db_file))
sql = 'SELECT * from test'
        'where Grades in (76,77,78)'
sales_data_df = pd.read_sql(sql, engine)
sales_data_df
```

This code creates a link to the database file called gradedata.db and runs a query against it. It then loads the data resulting from that query into the dataframe called sales_data_df. If you don't know the names of the tables in a sqlite database, you can find out by changing the SQL statement to that shown in Listing 2-18.

Listing 2-18. Finding the Table Names

```
sql = "select name from sqlite_master"
    "where type = 'table';"
```

Once you know the name of a table you wish to view (let's say it was test), if you want to know the names of the fields in that table, you can change your SQL statement to that shown in Listing 2-19.

Listing 2-19. A Basic Query

```
sql = "select * from test;"
```

Then, once you run sales_data_df.head() on the dataframe, you will be able to see the fields as headers at the top of each column.

As always, if you need more information about the command, you can run the code shown in Listing 2-20.

Listing 2-20. Get Help on read_sql

```
sales_data_df.read_sql?
```

Your Turn

Can you load data from the datasets/salesdata.db database?

Saving Data to SQL

See Listing 2-21 for an example of how to do this.

Listing 2-21. Create Dataset to Save

```
import pandas as pd
names = ['Bob','Jessica','Mary','John','Mel']
grades = [76,95,77,78,99]
GradeList = zip(names,grades)
df = pd.DataFrame(data = GradeList,
    columns=['Names', 'Grades'])
df
```

To export it to SQL, we can use the code shown in Listing 2-22.

Listing 2-22. Export Dataframe to sqlite

```
import os
import sqlite3 as lite
db_filename = r'mydb.db'
con = lite.connect(db_filename)
df.to_sql('mytable',
    con,
    flavor='sqlite',
    schema=None,
    if_exists='replace',
index=True,
index_label=None,
chunksize=None,
dtype=None)
con.close()
```

- Line 14: `mydb.db` is the path and name of the sqlite database you wish to use.

- Line 18: `mytable` is the name of the table in the database.

As always, if you need more information about the command, you can run the code shown in Listing 2-23.

Listing 2-23. Get Help on to_sql

```
df.to_sql?
```

Your Turn

This might be a little tricky, but can you create a sqlite table that contains the data found in `datasets/gradedata.csv`?

Random Numbers and Creating Random Data

Normally, you will use the techniques in this guide with datasets of real data. However, sometimes you will need to create random values.

Say we wanted to make a random list of baby names. We could get started as shown in Listing 2-24.

Listing 2-24. Getting Started

```
import pandas as pd
from numpy import random
from numpy.random import randint
names = ['Bob','Jessica','Mary','John','Mel']
```

First, we import our libraries as usual. In the last line, we create a list of the names we will randomly select from.

Next, we add the code shown in Listing 2-25.

Listing 2-25. Seeding Random Generator

```
random.seed(500)
```

This seeds the random number generator. If you use the same seed, you will get the same "random" numbers.

What we will try to do is this:

1. `randint(low=0,high=len(names))`

    Generates a random integer between zero and the length of the list names.

2. `names[n]`

    Selects the name where its index is equal to n.

3. `for i in range(n)`

    Loops until i is equal to n, i.e., 1,2,3,....*n*.

4. `random_names =`

    Selects a random name from the name list and does this n times.

We will do all of this in the code shown in Listing 2-26.

Listing 2-26. Selecting 1000 Random Names

```
randnames = []
for i in range(1000):
    name = names[randint(low=0,high=len(names))]
    randnames.append(name)
```

Now we have a list of 1000 random names saved in our random_names variable. Let's create a list of 1000 random numbers from 0 to 1000 (Listing 2-27).

Listing 2-27. Selecting 1000 Random Numbers

```
births = []
for i in range(1000):
    births.append(randint(low=0, high=1000))
```

And, finally, zip the two lists together and create the dataframe (Listing 2-28).

Listing 2-28. Creating Dataset from the Lists of Random Names and Numbers

```
BabyDataSet2 = list(zip(randnames,births))
df = pd.DataFrame(data = BabyDataSet2,
        columns=['Names', 'Births'])
df
```

Your Turn

Create a dataframe called parkingtickets with 250 rows containing a name and a number between 1 and 25.

CHAPTER 3

Preparing Data Is Half the Battle

The second step of data analysis is cleaning the data. Getting data ready for analytical tools can be a difficult task. Python and its libraries try to make it as easy as possible.

With just a few lines of code, you will be able to get your data ready for analysis. You will be able to

- clean the data;

- create new variables; and

- organize the data.

Cleaning Data

To be useful for most analytical tasks, data must be clean. This means it should be consistent, relevant, and standardized. In this chapter, you will learn how to

- remove outliers;

- remove inappropriate values;

- remove duplicates;

© A.J. Henley and Dave Wolf 2018
A.J. Henley and D. Wolf, *Learn Data Analysis with Python*,
https://doi.org/10.1007/978-1-4842-3486-0_3

- remove punctuation;

- remove whitespace;

- standardize dates; and

- standardize text.

Calculating and Removing Outliers

Assume you are collecting data on the people you went to high school with. What if you went to high school with Bill Gates? Now, even though the person with the second-highest net worth is only worth $1.5 million, the average of your entire class is pushed up by the billionaire at the top. Finding the outliers allows you to remove the values that are so high or so low that they skew the overall view of the data.

We cover two main ways of detecting outliers:

1. **Standard Deviations**: If the data is normally distributed, then 95 percent of the data is within 1.96 standard deviations of the mean. So we can drop the values either above or below that range.

2. **Interquartile Range (IQR)**: The IQR is the difference between the 25 percent quantile and the 75 percent quantile. Any values that are either lower than Q1 - 1.5 x IQR or greater than Q3 + 1.5 x IQR are treated as outliers and removed.

Let's see what these look like (Listings 3-1 and 3-2).

Listing 3-1. Method 1: Standard Deviation

```
import pandas as pd
Location = "datasets/gradedata.csv"
df = pd.read_csv(Location)
meangrade = df['grade'].mean()
```

```
stdgrade = df['grade'].std()
toprange = meangrade + stdgrade * 1.96
botrange = meangrade - stdgrade * 1.96
copydf = df
copydf = copydf.drop(copydf[copydf['grade']
        > toprange].index)
copydf = copydf.drop(copydf[copydf['grade']
        < botrange].index)
copydf
```

- Line 6: Here we calculate the upper range equal to 1.96 times the standard deviation plus the mean.

- Line 7: Here we calculate the lower range equal to 1.96 times the standard deviation subtracted from the mean.

- Line 9: Here we drop the rows where the grade is higher than the toprange.

- Line 11: Here we drop the rows where the grade is lower than the botrange.

Listing 3-2. Method 2: Interquartile Range

```
import pandas as pd
Location = "datasets/gradedata.csv"
df = pd.read_csv(Location)
q1 = df['grade'].quantile(.25)
q3 = df['grade'].quantile(.75)
iqr = q3-q1
toprange = q3 + iqr * 1.5
botrange = q1 - iqr * 1.5
copydf = df
```

```
copydf = copydf.drop(copydf[copydf['grade']
        > toprange].index)
copydf = copydf.drop(copydf[copydf['grade']
        < botrange].index)
copydf
```

- Line 9: Here we calculate the upper boundary = the third quartile + 1.5 * the IQR.

- Line 10: Here we calculate the lower boundary = the first quartile - 1.5 * the IQR.

- Line 13: Here we drop the rows where the grade is higher than the toprange.

- Line 14: Here we drop the rows where the grade is lower than the botrange.

Your Turn

Load the dataset datasets/outlierdata.csv. Can you remove the outliers? Try it with both methods.

Missing Data in Pandas Dataframes

One of the most annoying things about working with large datasets is finding the missing datum. It can make it impossible or unpredictable to compute most aggregate statistics or to generate pivot tables. If you look for missing data points in a 50-row dataset it is fairly easy. However, if you try to find a missing data point in a 500,000-row dataset it can be much tougher.

Python's pandas library has functions to help you find, delete, or change missing data (Listing 3-3).

Listing 3-3. Creating Dataframe with Missing Data

```
import pandas as pd
df = pd.read_csv("datasets/gradedatamissing.csv")
df.head()
```

The preceding code loads a legitimate dataset that includes rows with missing data. We can use the resulting dataframe to practice dealing with missing data.

To drop all the rows with missing (NaN) data, use the code shown in Listing 3-4.

Listing 3-4. Drop Rows with Missing Data

```
df_no_missing = df.dropna()
df_no_missing
```

To add a column filled with empty values, use the code in Listing 3-5.

Listing 3-5. Add a Column with Empty Values

```
import numpy as np
df['newcol'] = np.nan
df.head()
```

To drop any columns that contain nothing but empty values, see Listing 3-6.

Listing 3-6. Drop Completely Empty Columns

```
df.dropna(axis=1, how='all')
```

To replace all empty values with zero, see Listing 3-7.

Listing 3-7. Replace Empty Cells with 0

```
df.fillna(0)
```

To fill in missing grades with the mean value of grade, see Listing 3-8.

Listing 3-8. Replace Empty Cells with Average of Column

```
df["grade"].fillna(df["grade"].mean(), inplace=True)
```

Note, `inplace=True` means that the changes are saved to the dataframe right away.

To fill in missing grades with each gender's mean value of grade, see Listing 3-9.

Listing 3-9. It's Complicated

```
df["grade"].fillna(df.groupby("gender")
    ["grade"].transform("mean"), inplace=True)
```

We can also select some rows but ignore the ones with missing data points. To select the rows of df where age is not NaN and gender is not NaN, see Listing 3-10.

Listing 3-10. Selecting Rows with No Missing Age or Gender

```
df[df['age'].notnull() & df['gender'].notnull()]
```

Your Turn

Load the dataset `datasets/missinggrade.csv`. Your mission, if you choose to accept it, is to delete rows with missing grades and to replace the missing values in hours of exercise by the mean value for that gender.

Filtering Inappropriate Values

Sometimes, if you are working with data you didn't collect yourself, you need to worry about whether the data is accurate. Heck, sometimes you need to worry about that even if you did collect it yourself! It can be difficult to check the veracity of each and every data point, but it is quite easy to check if the data is appropriate.

Python's pandas library has the ability to filter out the bad values (see Listing 3-11).

Listing 3-11. Creating Dataset

```
import pandas as pd

names = ['Bob','Jessica','Mary','John','Mel']
grades = [76,-2,77,78,101]

GradeList = zip(names,grades)
df = pd.DataFrame(data = GradeList,
    columns=['Names', 'Grades'])
df
```

To eliminate all the rows where the grades are too high, see Listing 3-12.

Listing 3-12. Filtering Out Impossible Grades

```
df.loc[df['Grades'] <= 100]
```

To change the out-of-bound values to the maximum or minimum allowed value, we can use the code seen in Listing 3-13.

Listing 3-13. Changing Impossible Grades

```
df.loc[(df['Grades'] >= 100,'Grades')] = 100
```

Your Turn

Using the dataset from this section, can you replace all the subzero grades with a grade of zero?

Finding Duplicate Rows

Another thing you need to worry about if you are using someone else's data is whether any data is duplicated. (Did the same data get reported twice, or recorded twice, or just copied and pasted?) Heck, sometimes you need to worry about that even if you did collect it yourself! It can be difficult to check the veracity of each and every data point, but it is quite easy to check if the data is duplicated.

Python's pandas library has a function for finding not only duplicated rows, but also the unique rows (Listing 3-14).

Listing 3-14. Creating Dataset with Duplicates

```
import pandas as pd
names = ['Jan','John','Bob','Jan','Mary','Jon','Mel','Mel']
grades = [95,78,76,95,77,78,99,100]
GradeList = zip(names,grades)
df = pd.DataFrame(data = GradeList,
        columns=['Names', 'Grades'])
df
```

To indicate the duplicate rows, we can simply run the code seen in Listing 3-15.

Listing 3-15. Displaying Only Duplicates in the Dataframe

```
df.duplicated()
```

To show the dataset without duplicates, we can run the code seen in Listing 3-16.

Listing 3-16. Displaying Dataset without Duplicates

```
df.drop_duplicates()
```

You might be asking, "What if the entire row isn't duplicated, but I still know it's a duplicate?" This can happen if someone does your survey or retakes an exam again, so the name is the same, but the observation is different. In this case, where we know that a duplicate name means a duplicate entry, we can use the code seen in Listing 3-17.

Listing 3-17. Drop Rows with Duplicate Names, Keeping the Last Observation

```
df.drop_duplicates(['Names'], keep='last')
```

Your Turn

Load the dataset datasets/dupedata.csv. We figure people with the same address are duplicates. Can you drop the duplicated rows while keeping the first?

Removing Punctuation from Column Contents

Whether in a phone number or an address, you will often find unwanted punctuation in your data. Let's load some data to see how to address that (Listing 3-18).

Listing 3-18. Loading Dataframe with Data from CSV File

```
import pandas as pd
Location = "datasets/gradedata.csv"
## To add headers as we load the data...
df = pd.read_csv(Location)
df.head()
```

To remove the unwanted punctuation, we create a function that returns all characters that aren't punctuation, and them we apply that function to our dataframe (Listing 3-19).

Listing 3-19. Stripping Punctuation from the Address Column

```python
import string
exclude = set(string.punctuation)
def remove_punctuation(x):
    try:
        x = ''.join(ch for ch in x if ch not in exclude)
    except:
        pass
    return x
df.address = df.address.apply(remove_punctuation)
df
```

Removing Whitespace from Column Contents

Listing 3-20. Loading Dataframe with Data from CSV File

```python
import pandas as pd
Location = "datasets/gradedata.csv"
## To add headers as we load the data...
df = pd.read_csv(Location)
df.head()
```

To remove the whitespace, we create a function that returns all characters that aren't punctuation, and them we apply that function to our dataframe (Listing 3-21).

Listing 3-21. Stripping Whitespace from the Address Column

```python
def remove_whitespace(x):
    try:
        x = ''.join(x.split())
    except:
```

```
        pass
    return x
df.address = df.address.apply(remove_whitespace)
df
```

Standardizing Dates

One of the problems with consolidating data from different sources is that different people and different systems can record dates differently. Maybe they use 01/03/1980 or they use 01/03/80 or even they use 1980/01/03. Even though they all refer to January 3, 1980, analysis tools may not recognize them all as dates if you are switching back and forth between the different formats in the same column (Listing 3-22).

Listing 3-22. Creating Dataframe with Different Date Formats

```
import pandas as pd
names = ['Bob','Jessica','Mary','John','Mel']
grades = [76,95,77,78,99]
bsdegrees = [1,1,0,0,1]
msdegrees = [2,1,0,0,0]
phddegrees = [0,1,0,0,0]
bdates = ['1/1/1945','10/21/76','3/3/90',
        '04/30/1901','1963-09-01']
GradeList = zip(names,grades,bsdegrees,msdegrees,
        phddegrees,bdates)
columns=['Names','Grades','BS','MS','PhD',"bdates"]
df = pd.DataFrame(data = GradeList, columns=columns)
df
```

Listing 3-23 shows a function that standardizes dates to single format.

Listing 3-23. Function to Standardize Dates

```
from time import strftime
from datetime import datetime
def standardize_date(thedate):
    formatted_date = ""
    thedate = str(thedate)
    if not thedate or thedate.lower() == "missing"
                or thedate == "nan":
        formatted_date = "MISSING"
    if the_date.lower().find('x') != -1:
        formatted_date = "Incomplete"
    if the_date[0:2] == "00":
        formatted_date = thedate.replace("00", "19")
    try:
        formatted_date = str(datetime.strptime(
        thedate,'%m/%d/%y')
.strftime('%m/%d/%y'))
    except:
        pass
    try:
        formatted_date = str(datetime.strptime(
thedate, '%m/%d/%Y')
.strftime('%m/%d/%y'))
    except:
        pass
    try:
        if int(the_date[0:4]) < 1900:
            formatted_date = "Incomplete"
        else:
            formatted_date = str(datetime.strptime(
            thedate, '%Y-%m-%d')
```

```
.strftime('%m/%d/%y'))
    except:
        pass
    return formatted_date
```

Now that we have this function, we can apply it to the `birthdates` column on our dataframe (Listing 3-24).

Listing 3-24. Applying Date Standardization to Birthdate Column

```
df.bdates = df.bdates.apply(standardize_date)
df
```

Standardizing Text like SSNs, Phone Numbers, and Zip Codes

One of the problems with consolidating data from different sources is that different people and different systems can record certain data like Social Security numbers, phone numbers, and zip codes differently. Maybe they use hyphens in those numbers, and maybe they don't. This section quickly covers how to standardize how these types of data are stored (see Listing 3-25).

Listing 3-25. Creating Dataframe with SSNs

```
import pandas as pd
names = ['Bob','Jessica','Mary','John','Mel']
grades = [76,95,77,78,99]
bsdegrees = [1,1,0,0,1]
msdegrees = [2,1,0,0,0]
phddegrees = [0,1,0,0,0]
ssns = ['867-53-0909','333-22-4444','123-12-1234',
        '777-93-9311','123-12-1423']
```

```
GradeList = zip(names,grades,bsdegrees,msdegrees,
        phddegrees,ssns)
columns=['Names','Grades','BS','MS','PhD',"ssn"]
df = pd.DataFrame(data = GradeList, columns=columns)
df
```

The code in Listing 3-26 creates a function that standardizes the SSNs and applies it to our ssn column.

Listing 3-26. Remove Hyphens from SSNs and Add Leading Zeros if Necessary

```
def right(s, amount):
    return s[-amount]
def standardize_ssn(ssn):
    try:
        ssn = ssn.replace("-","")
        ssn = "".join(ssn.split())
        if len(ssn)<9 and ssn != 'Missing':
            ssn="000000000" + ssn
            ssn=right(ssn,9)
    except:
        pass
    return ssn
df.ssn = df.ssn.apply(standardize_ssn)
df
```

Creating New Variables

Once the data is free of errors, you need to set up the variables that will directly answer your questions. It's a rare dataset in which every question you need answered is directly addressed by a variable. So, you may need to do a lot of recoding and computing of variables to get exactly the dataset that you need.

Examples include the following:

- Creating bins (like converting numeric grades to letter grades or ranges of dates into Q1, Q2, etc.)

- Creating a column that ranks the values in another column

- Creating a column to indicate that another value has reached a threshold (passing or failing, Dean's list, etc.)

- Converting string categories to numbers (for regression or correlation)

Binning Data

Sometimes, you will have discrete data that you need to group into bins. (Think: converting numeric grades to letter grades.) In this lesson, we will learn about binning (Listing 3-27).

Listing 3-27. Loading the Dataset from CSV

```
import pandas as pd
Location = "datasets/gradedata.csv"
df = pd.read_csv(Location)
df.head()
```

Now that the data is loaded, we need to define the bins and group names (Listing 3-28).

Listing 3-28. Define Bins as 0 to 60, 60 to 70, 70 to 80, 80 to 90, 90 to 100

```
# Create the bin dividers
bins = [0, 60, 70, 80, 90, 100]
# Create names for the four groups
group_names = ['F', 'D', 'C', 'B', 'A']
```

Notice that there is one more bin value than there are group_names. This is because there needs to be a top and bottom limit for each bin.

Listing 3-29. Cut Grades

```
df['lettergrade'] = pd.cut(df['grade'], bins,
        labels=group_names)
df
```

Listing 3-29 categorizes the column grade based on the bins list and labels the values using the group_names list.

And if we want to count the number of observations for each category, we can do that too (Listing 3-30).

Listing 3-30. Count Number of Observations

```
pd.value_counts(df['lettergrade'])
```

Your Turn

Recreate the dataframe from this section and create a column classifying the row as pass or fail. This is for a master's program that requires a grade of 80 or above for a student to pass.

Applying Functions to Groups, Bins, and Columns

The number one reason I use Python to analyze data is to handle datasets larger than a million rows. The number two reason is the ease of applying functions to my data.

To see this, first we need to load up some data (Listing 3-31).

Listing 3-31. Loading a Dataframe from a CSV File

```
import pandas as pd
Location = "datasets/gradedata.csv"
df = pd.read_csv(Location)
df.head()
```

Then, we use binning to divide the data into letter grades (Listing 3-32).

Listing 3-32. Using Bins

```
# Create the bin dividers
bins = [0, 60, 70, 80, 90, 100]
# Create names for the four groups
group_names = ['F', 'D', 'C', 'B', 'A']
df['letterGrades'] = pd.cut(df['grade'],
        bins, labels=group_names)
df.head()
```

To find the average hours of study by letter grade, we apply our functions to the binned column (Listing 3-33).

Listing 3-33. Applying Function to Newly Created Bin

```
df.groupby('letterGrades')['hours'].mean()
```

Applying a function to a column looks like Listing 3-34.

Listing 3-34. Applying a Function to a Column

```
# Applying the integer function to the grade column
df['grade'] = df['grade'] = df['grade']
.apply(lambda x: int(x))
df.head()
```

- Line 1: Let's get an integer value for each grade in the dataframe.

Applying a function to a group can be seen in Listing 3-35.

Listing 3-35. Applying a Function to a Group

```
gender_preScore = df['grade'].groupby(df['gender'])
gender_preScore.mean()
```

- Line 1: Create a grouping object. In other words, create an object that represents that particular grouping. In this case, we group grades by the gender.

- Line 2: Display the mean value of each regiment's pre-test score.

Your Turn

Import the datasets/gradedata.csv file and create a new binned column of the 'status' as either passing (> 70) or failing (<=70). Then, compute the mean hours of exercise of the female students with a 'status' of passing.

Ranking Rows of Data

It is relatively easy to find the row with the maximum value or the minimum value, but sometimes you want to find the rows with the 50 highest or the 100 lowest values for a particular column. This is when you need ranking (Listing 3-36).

Listing 3-36. Load Data from CSV

```
import pandas as pd
Location = "datasets/gradedata.csv"
df = pd.read_csv(Location)
df.head()
```

If we want to find the rows with the lowest grades, we will need to rank all rows in ascending order by grade. Listing 3-37 shows the code to create a new column that is the rank of the value of grade in ascending order.

Listing 3-37. Create Column with Ranking by Grade

```
df['graderanked'] = df['grade'].rank(ascending=1)
df.tail()
```

So, if we just wanted to see the students with the 20 lowest grades, we would use the code in Listing 3-38.

Listing 3-38. Bottom 20 Students

```
df[df['graderanked'] < 21]
```

And, to see them in order, we need to use the code in Listing 3-39.

Listing 3-39. Bottom 6 Students in Order

```
df[df['graderanked'] < 6].sort_values('graderanked')
```

Your Turn

Can you find the 50 students with the most hours of study per week?

Create a Column Based on a Conditional

Sometimes, you need to classify a row of data by the values in one or more columns, such as identifying those students who are passing or failing by whether their grade is above or below 70. In this section, we will learn how to do this (Listing 3-40).

Listing 3-40. Load Data from CSV

```
import pandas as pd
Location = "datasets/gradedata.csv"
df = pd.read_csv(Location)
df.head()
```

Now, let us say we want a column indicating whether students are failing or not (Listing 3-41).

Listing 3-41. Create Yes/No isFailing Column

```
import numpy as np
df['isFailing'] = np.where(df['grade']<70,
'yes', 'no')
df.tail(10)
```

> Line 1: We import the numpy library
>
> Line 2: Create a new column called df.failing where the value is yes if df.grade is less than 70 and no if not.

If instead we needed a column indicating who the male students were with failing scores, we could use the code in Listing 3-42.

Listing 3-42. Create Yes/No isFailingMale Column

```
df['isFailingMale'] = np.where((df['grade']<70)
        & (df['gender'] == 'male'),'yes', 'no')
df.tail(10)
```

Your Turn

Can you create a column for timemgmt that shows busy if a student exercises more than three hours per week AND studies more than seventeen hours per week?

Making New Columns Using Functions

Much of what I used to use Excel to do (and what I now use Python for) is to create new columns based on an existing one. So, using the following data (Listing 3-43), let's see how we would do this.

Listing 3-43. Load Data from CSV

```
import pandas as pd
Location = "datasets/gradedata.csv"
df = pd.read_csv(Location)
df.head()
```

To create a single column to contain the full name of each student, we first create a function to create a single string from two strings (Listing 3-44).

Listing 3-44. Create Function to Generate Full Name

```
def singlename(fn, ln):
    return fn + " " + ln
```

Now, if you test that function, you will see that it works perfectly well concatenating Adam and Smith into Adam Smith. However, we can also use it with column selectors to create a new column using our fname and lname columns (Listing 3-45).

Listing 3-45. Create Column to Hold the Full Name

```
df['fullname'] = singlename(df['fname'],df['lname'])
```

This code creates a column called fullname that concatenates the first and last name.

Your Turn

Can you create a column called total time that adds together the hours of study per week and the hours of exercise per week?

Converting String Categories to Numeric Variables

Why do I need to convert string categories to numeric variables? Many analytical tools won't work on text, but if you convert those values to numbers it makes things much simpler (Listing 3-46).

Listing 3-46. Load Data from CSV

```
import pandas as pd
Location = "datasets/gradedata.csv"
df = pd.read_csv(Location)
df.head()
```

Method 1: Convert single column to hold numeric variables (Listing 3-47).

Listing 3-47. Function to Convert Gender to Number

```
def score_to_numeric(x):
    if x=='female':
        return 1
    if x=='male':
        return 0
```

Now, run that method on your column (Listing 3-48).

Listing 3-48. Apply score_to_numeric Function to Gender

```
df['gender_val'] = df['gender'].apply(score_to_numeric)
df.tail()
```

Method 2: Create individual Boolean columns (Listing 3-49).

Listing 3-49. Create Boolean Columns Based on Gender Column

```
df_gender = pd.get_dummies(df['gender'])
df_gender.tail()
```

Join columns to original dataset (Listing 3-50).

Listing 3-50. Add New Columns to Original Dataframe

```
# Join the dummy variables to the main dataframe
df_new = pd.concat([df, df_gender], axis=1)
df_new.tail()
# or
# Alterative for joining the new columns
df_new = df.join(df_gender)
df_new.tail()
```

41

Your Turn

Using `datasets/gradesdatawithyear.csv`, can you create a numeric column to replace statuses of freshman through senior with the numerals 1 through 4?

Organizing the Data

Both original and newly created variables need to be formatted correctly for two reasons.

First, so our analysis tools work with them correctly. Failing to format a missing value code or a dummy variable correctly will have major consequences for your data analysis.

Second, it's much faster to run the analysis and interpret results if you don't have to keep looking up which variable Q156 is.

Examples include the following:

- Removing columns that aren't needed

- Changing column names

- Changing column names to lower case

- Formatting date variables as dates, and so forth.

Removing and Adding Columns

Sometimes we need to adjust the data. Either something is left out that should have been included or something was left in that should have been removed. So, let's start with the dataset in Listing 3-51.

Listing 3-51. Creating Starting Dataset

```
import pandas as pd
names = ['Bob','Jessica','Mary','John','Mel']
```

```
grades = [76,95,77,78,99]
bsdegrees = [1,1,0,0,1]
msdegrees = [2,1,0,0,0]
phddegrees = [0,1,0,0,0]
GradeList = zip(names,grades,bsdegrees,msdegrees,
        phddegrees)
columns=['Names','Grades','BS','MS','PhD']
df = pd.DataFrame(data = GradeList, columns=columns)
df
```

We can drop a column by simply adding the code in Listing 3-52.

Listing 3-52. Dropping a Column

```
df.drop('PhD', axis=1)
```

With axis=1 telling drop that we want to drop a column (1) and not a row (0).

We can add a column filled with zeros by setting the new column name to be equal to a 0 (Listing 3-53).

Listing 3-53. Creating a New Column Filled with Zeros

```
df['HighSchool']=0
```

If, however, you want to set the new columns to equal null values, you can do that too (Listing 3-54).

Listing 3-54. Creating a New Column Filled with Null Values

```
df['PreSchool'] = np.nan
```

Now, adding a column with values is not that hard. We create a series and set the column equal to the series (Listing 3-55).

Listing 3-55. Creating a New Column Filled with Values

```
d = ([0,1,0,1,0])
s = pd.Series(d, index= df.index)
df['DriversLicense'] = s
df
```

Your Turn

Can you remove the bs, ms, and phd degree columns?

Can you add a Hogwarts Magic Degree column? Everyone but Jessica has one; does that make it harder? No? Then I have to be sure to stump you next time.

Selecting Columns

You will need to make subselections of your data occasionally, especially if your dataset has tons of columns. Here, we learn how to create a dataframe that includes only some of our columns (Listing 3-56).

Listing 3-56. Load Data from CSV

```
import pandas as pd
Location = "datasets/gradedata.csv"
df = pd.read_csv(Location)
df.head()
```

Now, to select a column of data, we specify the column name (Listing 3-57).

Listing 3-57. Selecting a Column into a List

```
df['fname']
```

But if you run that code you only get the data in the column (notice the header is missing). That is because this doesn't return a dataframe; it returns a list. To return a dataframe when selecting a column, we need to specify it (Listing 3-58).

Listing 3-58. Selecting a Column into a Dataframe

```
df[['fname']]
```

To return multiple columns, we use code like that in Listing 3-59.

Listing 3-59. Selecting Multiple Columns into a Dataframe

```
df[['fname','age','grade']]
```

And, of course, if we want to create a dataframe with that subset of columns, we can copy it to another variable (Listing 3-60).

Listing 3-60. Creating New Dataframe from Your Selection

```
df2 = df[['fname','age','grade']]
df2.head()
```

Your Turn

We need to create a mailing list. Can you create a new dataframe by selecting the first name, last name, and address fields?

Change Column Name

Sometimes you need change the names of your columns. With pandas, it's easy to do. First, you load your data (Listing 3-61).

45

Listing 3-61. Load Dataset from CSV

```
import pandas as pd

Location = "datasets/gradedata.csv"
df = pd.read_csv(Location)
df.head()
```

But, when we look at the header, we aren't crazy about the column names—or it doesn't have any.

It is simple to change the column headers (Listing 3-62).

Listing 3-62. Change All Headers

```
df.columns = ['FirstName', 'LastName', 'Gender',
        'Age', 'HoursExercisePerWeek',
        'HoursStudyPerWeek', 'LetterGrade',
        'StreetAddress']
df.head()
```

Or, if you just wanted to change one or two values, you can load the list of headers (Listing 3-63).

Listing 3-63. Load List of Headers into a Temp Variable

```
headers = list(df.columns.values)
```

Once the headers are loaded, you can change a few (Listing 3-64).

Listing 3-64. Changing Headers

```
headers[0] = 'FName'
headers[1] = 'LName'
df.columns = headers
df.head()
```

Your Turn

Can you change the age column name to years?

Setting Column Names to Lower Case

It may not be the biggest problem in the world, but sometimes I need to convert all the column names to lowercase (or uppercase, for that matter). This lesson will cover how to do that (Listing 3-65).

Listing 3-65. Load Data from CSV

```
import pandas as pd
Location = "datasets/gradedata.csv"
df = pd.read_csv(Location)
df.head()
```

Once you have the data, there are two quick ways to cast all the column headers to lowercase (Listing 3-66).

Listing 3-66. Casting All Headers to Lowercase

```
# method 1
df.columns = map(str.lower, df.columns)
# method 2
df.columns = [x.lower() for x in df.columns]
```

Your Turn

Can you figure out how to make all the column headers all uppercase?

47

Finding Matching Rows

Of course, you don't always want to compute using the entire dataset.
Sometimes you want to work with just a subset of your data. In this lesson,
we find out how to do that (Listing 3-67).

Listing 3-67. Creating Dataset

```
import pandas as pd
names = ['Bob','Jessica','Mary','John','Mel']
grades = [76,95,77,78,99]
GradeList = zip(names,grades)
df = pd.DataFrame(data = GradeList,
        columns=['Names', 'Grades'])
df
```

To find all the rows that contain the word *Mel*, use the code shown in
Listing 3-68 in a new cell.

Listing 3-68. Filtering Rows

```
df['Names'].str.contains('Mel')
```

After executing that line of Python, you will see a list of Boolean
values—True for the lines that match our query and False for that ones that
don't.

We can make our answer shorter by adding .any. This will just display a
single True if any line matches and False if none of them do (Listing 3-69).

Listing 3-69. Check if Any Rows Match

```
# check if any row matches
df['Names'].str.contains('Mel').any()
```

Alternatively, you can add .all. This will just display a single True if all of the lines match and False if at least one of them does not (Listing 3-70).

Listing 3-70. Check if All Rows Match

```
# check if all rows match
df['Names'].str.contains('Mel').all()
```

We can also use this along with the .loc (locate) function to show just the rows that match certain criteria (Listing 3-71).

Listing 3-71. Show the Rows that Match

```
# Find the rows that match a criteria like this
df.loc[df['Names'].str.contains('Mel')==True]
# or even like this...
df.loc[df['Grades']==0]
```

Your Turn

Can you find all the people who have at least one MS degree in the following data (Listing 3-72)?

Listing 3-72. Starting Dataset

```
import pandas as pd
names = ['Bob','Jessi','Mary','John','Mel','Sam',
        'Cathy','Hank','Lloyd']
grades = [76,95,77,78,99,84,79,100,73]
bsdegrees = [1,1,0,0,1,1,1,0,1]
msdegrees = [2,1,0,0,0,1,1,0,0]
phddegrees = [0,1,0,0,0,2,1,0,0]
GradeList = zip(names,grades,bsdegrees,msdegrees,
        phddegrees)
```

```
df = pd.DataFrame(data = GradeList, columns=['Name','Grade','BS',
'MS','PhD'])
df
```

Filter Rows Based on Conditions

Listing 3-73. Load Data from CSV

```
import pandas as pd
Location = "datasets/gradedata.csv"
df = pd.read_csv(Location)
df.head()
```

We can show one column of data (Listing 3-74).

Listing 3-74. One Column

```
df['grade'].head()
```

Or we can show two columns of data (Listing 3-75).

Listing 3-75. Two Columns

```
df[['age','grade']].head()
```

Or we can show the first two rows of data (Listing 3-76).

Listing 3-76. First Two Rows

```
df[:2]
```

To show all the rows where the grade is greater than 80, use the code in Listing 3-77.

Listing 3-77. All Rows with Grade > 80

```
df[df['grade'] > 80]
```

Using multiple conditions is a little trickier. So, if we wanted to get a list of all the students who scored higher than 99.9 and were male, we would need to use the code shown in Listing 3-78.

Listing 3-78. All Rows Where Men Scored > 99.9

```
df.ix[(df['grade'] > 99.9) &
    (df['gender'] == 'male') ]
```

If instead we wanted all the students who scored higher than 99 OR were female, we would need to use the code in Listing 3-79.

Listing 3-79. All Rows Where Women or Scored > 99

```
df.ix[(df['grade'] > 99) | (df['gender'] == 'female') ]
```

Your Turn

Can you show all the rows where the student was male, exercised less than two hours per week, and studied more than fifteen hours per week?

Selecting Rows Based on Conditions

Listing 3-80. Load Data from CSV

```
import pandas as pd
Location = "datasets/gradedata.csv"
df = pd.read_csv(Location)
df.head()
```

Listing 3-81. Method 1: Using Variables to Hold Attributes

```
female = df['gender'] == "female"
a_student = df['grade'] >= 90
df[female & a_student].head()
```

51

Line 1: We create a variable with TRUE if gender is female.

Line 2: We create a variable with TRUE if grade is greater than or equal to 90.

Line 3: This is where we select all cases where both gender is female and grade is greater than or equal to 90.

Listing 3-82. Method 2: Using Variable Attributes Directly

```
df[df['fname'].notnull() & (df['gender'] == "male")]
```

In Listing 3-82, we select all cases where the first name is not missing and gender is male.

Your Turn

Can you find all the rows where the student had four or more hours of exercise per week, seventeen or more hours of study, and still had a grade that was lower than 80?

Random Sampling Dataframe

This one is simple. Obviously, sometimes we have datasets that are too large and we need to take a subset, so let's start with some loaded data (Listing 3-83).

Listing 3-83. Load Dataset from CSV

```
import pandas as pd
import numpy as np
Location = "datasets/gradedata.csv"
df = pd.read_csv(Location)
df.tail()
```

To select just 100 rows randomly from that dataset, we can simply run the code shown in Listing 3-84.

Listing 3-84. Random Sample of 100 Rows from Dataframe

```
df.take(np.random.permutation(len(df))[:100])
```

Your Turn

Can you create a random sample of 500 rows from that dataset?

CHAPTER 4

Finding the Meaning

The third stage of data analysis is actually analyzing the data. Finding meaning within your data can be difficult without the right tools. In this section, we look at some of the tools available to the Python user.

With just a few lines of code, you will be able to do the following analysis:

- Compute descriptive statistics

- Correlation

- Linear regression

- Pivot tables

Computing Aggregate Statistics

As you may have seen in the last chapter, it is easy to get some summary statistics by using describe. Let's take a look at how we can find those values directly.

First, let's create some data (Listing 4-1).

Listing 4-1. Creating Dataset for Statistics

```
import pandas as pd
names = ['Bob','Jessica','Mary','John','Mel']
grades = [76,95,77,78,99]
```

```
GradeList = zip(names,grades)
df = pd.DataFrame(data=GradeList,
        columns=['Names','Grades'])
df
```

Once that is set up, we can take a look at some statistics (Listing 4-2).

Listing 4-2. Computing Aggregate Statistics

```
df['Grades'].count()  # number of values
df['Grades'].mean()   # arithmetic average
df['Grades'].std()    # standard deviation
df['Grades'].min()    # minimum
df['Grades'].max()    # maximum
df['Grades'].quantile(.25)  # first quartile
df['Grades'].quantile(.5)   # second quartile
df['Grades'].quantile(.75)  # third quartile
```

Note   If you tried to execute the previous code in one cell all at the same time, the only thing you would see is the output of the .quantile() function. You have to try them one by one. I simply grouped them all together for reference purposes. OK?

It is important to note that the mean is not the only measure of central tendency. See Listing 4-3 for other measures.

Listing 4-3. Other Measures of Central Tendency

```
# computes the arithmetic average of a column
# mean = dividing the sum by the number of values
df['Grades'].mean()
# finds the median of the values in a column
```

```
# median = the middle value if they are sorted in order
df['Grades'].median()
# finds the mode of the values in a column
# mode = the most common single value
df['Grades'].mode()
```

And if you need to compute standard deviation, you might also need variance (Listing 4-4).

Listing 4-4. Computing Variance

```
# computes the variance of the values in a column
df['Grades'].var()
```

Finally, you don't have to specify the column to compute the statistics. If you just run it against the whole dataframe, you will get the function to run on all applicable columns (Listing 4-5).

Listing 4-5. Computing Variance on All Numeric Columns

```
df.var()
```

Your Turn

Of course, in our dataset we only have one column. Try creating a dataframe and computing summary statistics using the dataset in Listing 4-6.

Listing 4-6. Starting Dataset

```
names = ['Bob','Jessica','Mary','John','Mel']
grades = [76,95,77,78,99]
bsdegrees = [1,1,0,0,1]
msdegrees = [2,1,0,0,0]
phddegrees = [0,1,0,0,0]
```

Computing Aggregate Statistics on Matching Rows

It is possible to compute descriptive statistics on just the rows that match certain criteria. First, let's create some data (Listing 4-7).

Listing 4-7. Creating Dataset

```
import pandas as pd
names = ['Bob','Jessica','Mary','John','Mel']
grades = [76,95,77,78,99]
bs = [1,1,0,0,1]
ms = [2,1,0,0,0]
phd = [0,1,0,0,0]
GradeList = zip(names,grades,bs,ms,phd)
df = pd.DataFrame(data=GradeList,
        columns=['Name','Grade','BS','MS','PhD'])
df
```

Ok, we have covered how to find the rows that match a set of criteria. We have also covered how to compute descriptive statistics, both all at once and one by one. If you put those two together, you will be able to find the statistics of the data that matches certain criteria.

So, to count the rows of the people without a PhD, use the code shown in Listing 4-8.

Listing 4-8. Code for Computing Aggregate Statistics

```
df.loc[df['PhD']==0].count()
```

You can use any of the aggregate statistics functions on individual columns in the same way. So, to find the average grade of those people without a PhD, use the code in Listing 4-9.

Listing 4-9. Computing Aggregate Statistics on a Particular Column

```
df.loc[df['PhD']==0]['Grade'].mean()
```

Your Turn

Using the data from Listing 4-10, what is the average grade for people with master's degrees?

Listing 4-10. Dataset for Exercise

```
import pandas as pd
names = ['Bob','Jessica','Mary','John',
        'Mel','Sam','Cathy','Henry','Lloyd']
grades = [76,95,77,78,99,84,79,100,73]
bs = [1,1,0,0,1,1,1,0,1]
ms = [2,1,0,0,0,1,1,0,0]
phd = [0,1,0,0,0,2,1,0,0]
GradeList = zip(names,grades,bs,ms,phd)
df = pd.DataFrame(data=GradeList,
        columns=['Names','Grades','BS','MS','PhD'])
df
```

Sorting Data

Generally, we get data in a random order, but need to use it in a completely different order. We can use the sort_values function to rearrange our data to our needs (Listing 4-11).

Listing 4-11. Loading Data from CSV

```
import pandas as pd
Location = "datasets/gradedata.csv"
df = pd.read_csv(Location)
df.head()
```

Sort the dataframe's rows by age, in descending order (Listing 4-12).

Listing 4-12. Sorting by Age, Descending

```
df = df.sort_values(by='age', ascending=0)
df.head()
```

Sort the dataframe's rows by hours of study and then by exercise, in ascending order (Listing 4-13).

Listing 4-13. Sorting by Hours of Study and Exercise, Ascending

```
df = df.sort_values(by=['grade', 'age'],
        ascending=[True, True])
df.head()
```

Your Turn

Can you sort the dataframe to order it by name, age, and then grade?

Correlation

Correlation is any of a broad class of statistical relationships involving dependence, though in common usage it most often refers to the extent to which two variables have a linear relationship with each other. Familiar examples of dependent phenomena include the correlation between the physical statures of parents and their offspring, and the correlation between the demand for a product and its price.

Basically, correlation measures how closely two variables move in the same direction. Tall parents have tall kids? Highly correlated. Wear lucky hat, but rarely win at cards? Very slightly correlated. As your standard of living goes up, your level of savings plummet? Highly negatively correlated.

Simple, right?

Well, computing correlation can be a little difficult by hand, but is totally simple in Python (Listing 4-14).

Listing 4-14. Running a Correlation

```
import pandas as pd
Location = "datasets/gradedata.csv"
df = pd.read_csv(Location)
df.head()
df.corr()
```

	Age	Exercise	Hours	Grade
Age	1.000000	-0.003643	-0.017467	-0.007580
Exercise	-0.003643	1.000000	0.021105	0.161286
Hours	-0.017467	0.021105	1.000000	0.801955
Grade	-0.007580	0.161286	0.801955	1.000000

The intersections with the highest absolute values are the columns that are the most correlated. Positive values are positively correlated, which means they go up together. Negative values are negatively correlated (as one goes up the other goes down). And, of course, each column is perfectly correlated with itself. As you can see, hours of study and grade are highly correlated.

Your Turn

Load the data from the code in Listing 4-15 and find the correlations.

Listing 4-15. Load Data from CSV

```
import pandas as pd
Location = "datasets/tamiami.csv"
```

Regression

In statistical modeling, regression analysis is a statistical process for estimating the relationships among variables. This is a fancy way of saying that we use regression to create an equation that explains the value of a dependent variable based on one or several independent variables. Let's get our data (Listing 4-16),

Listing 4-16. Load Data from CSV

```
import pandas as pd
Location = "datasets/gradedata.csv"
df = pd.read_csv(Location)
df.head()
```

Once we have that, we need to decide what columns we want to perform the regression on and which is the dependent variable. I want to try to predict the grade based on the age, hours of exercise, and hours of study (Listing 4-17).

Listing 4-17. First Regression

```
import statsmodels.formula.api as sm
result = sm.ols(
        formula='grade ~ age + exercise + hours',
        data=df).fit()
result.summary()
```

The formula format in line two is one that you need to learn and get comfortable editing. It shows the dependent variable on the left of the tilde (~) and the independent variables we want considered on the right.

If you look at the results you get from the summary, the R-squared represents the percentage of the variation in the data that can be accounted for by the regression. .664, or 66.4 percent, is good, but not great. The p-value (represented here by the value of P>|t|) represents the probability that the independent variable has no effect on the dependent variable. I like to keep my p-values less than 5 percent, so the only variable that stands out is the age with 59.5 percent. Let's rerun the regression, but leaving out the age (Listing 4-18).

Listing 4-18. Second Regression

```
import statsmodels.formula.api as sm
result = sm.ols(
        formula='grade ~ exercise + hours',
        data=df).fit()
result.summary()
```

Looking at our new results, we haven't changed our R-squared, but we have eliminated all our high p-values. So, we can now look at our coefficients, and we will end up with an equation that looks something like `grade = 1.916 * hours of study +.989 * hours of exercise + 58.5316`.

Your Turn

Create a new column where you convert gender to numeric values, like 1 for female and 0 for male. Can you now add gender to your regression? Does this improve your R-squared?

Regression without Intercept

Sometimes, your equation works better without an intercept. This can happen even though your *p*-values indicate otherwise. I always try it both ways, just as a matter of course, to see what the R-Squared is. To run your regression without an intercept, simply follow Listing 4-19.

Listing 4-19. Run Regression without Intercept

```
import pandas as pd
Location = "datasets/gradedata.csv"
df = pd.read_csv(Location)
df.head()
result = sm.ols(
        formula='grade ~ age + exercise + hours - 1', data=df).
fit()
result.summary()
```

Note that it is the - 1 at the end of the formula that tells Python that you wish to eliminate the intercept. If you look at the results, you can see we now have a much higher R-squared than we had in the last lesson, and we also have no *p*-values that cause us concern.

Your Turn

Try running these simple regressions with no intercept: 1. Tests for the relationship between just grade and age; 2. Tests for the relationship between just grade and exercise; and 3. Tests for the relationship between just grade and study.

If you had to pick just one, which one do you like best?

Basic Pivot Table

Pivot tables (or crosstabs) have revolutionized how Excel is used to do analysis. However, I like pivot tables in Python better than I do in Excel. Let's get some data (Listing 4-20).

Listing 4-20. Load Data from CSV

```
import pandas as pd
Location = "datasets/gradedata.csv"
df = pd.read_csv(Location)
df.head()
```

At its simplest, to get a pivot table we need a dataframe and an index (Listing 4-21).

Listing 4-21. Get Averages of All Numeric Columns Categorized by Gender

```
pd.pivot_table(df, index=['gender'])
```

As you can see, pivot_table is smart enough to assume that we want the averages of all the numeric columns. If we wanted to specify just one value, we could do that (Listing 4-22).

Listing 4-22. Average Grade by Gender

```
pd.pivot_table(df,
        values=['grade'],
        index=['gender'])
```

Gender	Grade
Female	82.7173
Male	82.3948

Now we see just the average grades, categorized by gender. If we wanted to, however, we could look at minimum hours of study (Listing 4-23).

Listing 4-23. Minimum Grade by Gender

```
pd.pivot_table(df,
        values=['grade'],
        index=['gender'],
        aggfunc='min')
```

Gender	Grade
Female	2
Male	0

We can also add other columns to the index. So, to view the maximum grade categorized by gender and age, we simply run the code in Listing 4-24.

Listing 4-24. Max Grade by Gender and Age

```
pd.pivot_table(df,
        index=['gender','age'],
        aggfunc='max',
        values=['hours'])
```

Gender	Age	Hours
Female	14	20
	15	20
	16	19
	17	20
	18	20
	19	20

Gender	Age	Hours
Male	14	19
	15	20
	16	20
	17	20
	18	20
	19	20

We can also have multiple value columns. So, to show the average grade and hours of study by gender, we can run the code in Listing 4-25.

Listing 4-25. Average Grade and Hours by Gender

```
pd.pivot_table(df,
        index=['gender'],
        aggfunc='mean',
        values=['grade','hours'])
```

Gender	Grade	Hours
Female	82.7173	10.932
Male	82.3948	11.045

We can also perform pivot tables on subsets of the data. First, select your data, then do a standard pivot on that selection. So, to show the average grade and hours of study by gender for students who are 17 years old, we can run the code in Listing 4-26.

Listing 4-26. Average Grade and Hours by Gender

```
df2 = df.loc[df['age'] == 17]
pd.pivot_table(df2,
```

```
index=['gender'],
aggfunc='mean',
values=['grade','hours'])
```

Gender	Grade	Hours
Female	83.599435	10.943503
Male	82.949721	11.268156

Finally, we can include totals on our Python pivot tables, as shown in Listing 4-27.

Listing 4-27. Average Grade and Hours by Gender

```
df2 = df.loc[df['age'] == 17]
pd.pivot_table(df2,
        index=['gender'],
        aggfunc='mean',
        values=['grade','hours'],
        margins='True')
```

Gender	Grade	Hours
Female	83.599435	10.943503
Male	82.949721	11.268156
All	83.272753	11.106742

Your Turn

Can you create a pivot table showing the average grade by gender of people who had more than two hours of exercise?

CHAPTER 5

Visualizing Data

Data Quality Report

When you have looked at enough datasets, you will develop a set of questions you want answered about the data to ascertain how good the dataset is. This following scripts combine to form a data quality report that I use to evaluate the datasets that I work with.

Listing 5-1. Load Dataset from CSV

```
# import the data
import pandas as pd
Location = "datasets\gradedata.csv"
df = pd.read_csv(Location)
df.head()
df.mode().transpose()
```

Listing 5-2. Finding Data Types of Each Column

```
data_types = pd.DataFrame(df.dtypes,
        columns=['Data Type'])
data_types
```

Listing 5-3. Counting Number of Missing Observations by Column

```
missing_data_counts = pd.DataFrame(df.isnull().sum(),
        columns=['Missing Values'])
missing_data_counts
```

Listing 5-4. Counting Number of Present Observations by Column

```
present_data_counts = pd.DataFrame(df.count(),
        columns=['Present Values'])
present_data_counts
```

Listing 5-5. Counting Number of Unique Observations by Column

```
unique_value_counts = pd.DataFrame(
        columns=['Unique Values'])
for v in list(df.columns.values):
        unique_value_counts.loc[v] = [df[v].nunique()]
unique_value_counts
```

Listing 5-6. Finding the Minimum Value for Each Column

```
minimum_values = pd.DataFrame(columns=[
        'Minimum Values'])
for v in list(df.columns.values):
        minimum_values.loc[v] = [df[v].min()]
minimum_values
```

Listing 5-7. Finding the Maximum Value for Each Column

```
maximum_values = pd.DataFrame(
        columns=['Maximum Values'])
for v in list(df.columns.values):
        maximum_values.loc[v] = [df[v].max()]
maximum_values
```

Listing 5-8. Joining All the Computed Lists into 1 Report

```
pd.concat([present_data_counts,
        missing_data_counts,
        unique_value_counts,
```

```
minimum_values,
maximum_values],
axis=1)
```

Your Turn

Can you create a data quality report for the datasets/tamiami.csv dataset?

Graph a Dataset: Line Plot

To create a simple line plot, input the code from Listing 5-9.

Listing 5-9. Line Plotting Your Dataset

```
import pandas as pd
names = ['Bob','Jessica','Mary','John','Mel']
grades = [76,83,77,78,95]
GradeList = zip(names,grades)
df = pd.DataFrame(data = GradeList,
        columns=['Names', 'Grades'])
%matplotlib inline
df.plot()
```

When you run it, you should see a graph that looks like Figure 5-1.

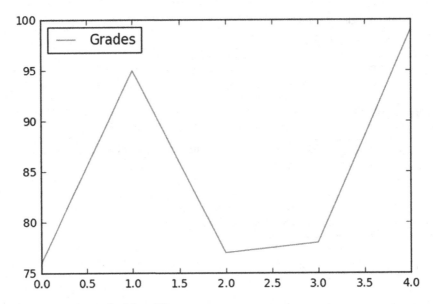

Figure 5-1. *Simple Line Plot*

Customizing the graph is easy, but you need to add the `matplotlib` library first.

Add the code in Listing 5-10 to what you did already.

Listing 5-10. Code to Plot a Customized Graph

```
import matplotlib.pyplot as plt
df.plot()
displayText = "my annotation"
xloc = 1
yloc = df['Grades'].max()
xtext = 8
ytext = 0
plt.annotate(displayText,
            xy=(xloc, yloc),
            xytext=(xtext,ytext),
            xycoords=('axes fraction', 'data'),
            textcoords='offset points')
```

Ok, the annotate command has pretty good documentation, located at
http://matplotlib.org/api/pyplot_api.html. But let's tear apart what
we typed:

> displayText: the text we want to show for this
> annotation

> xloc, yloc: the coordinates of the data point we
> want to annotate

> xtext, ytext: coordinates of where we want the
> text to appear using the coordinate system specified
> in textcoords

> xycoords: sets the coordinate system to use to find
> the data point; it can be set separately for x and y

> textcoords: sets the coordinate system to use to
> place the text

Finally, we can add an arrow linking the data point annotated to the
text annotation (Listing 5-11).

Listing 5-11. Code to Plot a Customized Graph

```
df.plot()
displayText = "my annotation"
xloc = 1
yloc = df['Grades'].max()
xtext = 8
ytext = -150
plt.annotate(displayText,
            xy=(xloc, yloc),
            arrowprops=dict(facecolor='black',
                            shrink=0.05),
```

```
        xytext=(xtext,ytext),
        xycoords=('axes fraction', 'data'),
        textcoords='offset points')
```

All we did is adjust the offset of the text so that there was enough room between the data and the annotation to actually see the arrow. We did this by changing the ytext value from 0 to -150. Then, we added the setting for the arrow.

More information about creating arrows can be found on the documentation page for annotate at http://matplotlib.org/users/ annotations_intro.html.

Your Turn

Take the same dataset we used in this example and add an annotation to Bob's 76 that says "Wow!"

Graph a Dataset: Bar Plot

To create a bar plot, input the code in Listing 5-12.

Listing 5-12. Bar Plotting Your Dataset

```
import matplotlib.pyplot as plt
import pandas as pd
names = ['Bob','Jessica','Mary','John','Mel']
status = ['Senior','Freshman','Sophomore','Senior',
          'Junior']
grades = [76,95,77,78,99]
GradeList = zip(names,grades)
```

```
df = pd.DataFrame(data = GradeList,
        columns=['Names', 'Grades'])
%matplotlib inline
df.plot(kind='bar')
```

Once you run it, you will get a simple bar plot, but the titles on the x-axis are the numbers 0–4.

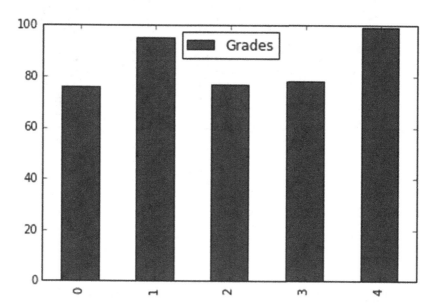

Figure 5-2. *Simple Bar Plot*

But if we convert the Names column into the index, we can improve the graph. So, first, we need to add the code in Listing 5-13.

Listing 5-13. Adding Code to Plot Your Dataset

```
df2 = df.set_index(df['Names'])
df2.plot(kind="bar")
```

We will then get a graph that looks like Figure 5-3.

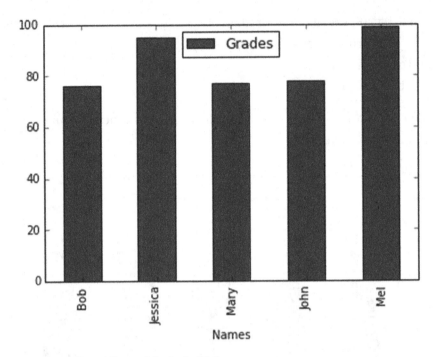

Figure 5-3. *Bar Plot with Axis Titles*

Your Turn

Can you change the code to create a bar plot where the status is the label?

Graph a Dataset: Box Plot

To create a box plot, input the code in Listing 5-14.

Listing 5-14. Box Plotting Your Dataset

```
import matplotlib.pyplot as plt
import pandas as pd
%matplotlib inline
names = ['Bob','Jessica','Mary','John','Mel']
```

```
grades = [76,95,77,78,99]
gender = ['Male','Female','Female','Male','Female']
status = ['Senior','Senior','Junior','Junior','Senior']
GradeList = zip(names,grades,gender)
df = pd.DataFrame(data = GradeList, columns=['Names', 'Grades',
'Gender'])
df.boxplot(column='Grades')
```

Once you run it, you will get a simple box plot.

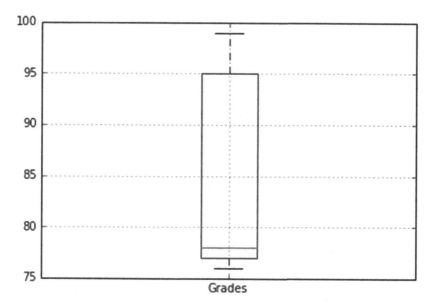

Figure 5-4. *Simple Box Plot*

Now, we can use a single command to create categorized graphs (in this case, categorized by gender). See Listing 5-15.

Listing 5-15. Adding Code to Categorize Your Box Plot

```
df.boxplot(by='Gender', column='Grades')
```

And we will then get a graph that looks like Figure 5-5. See Listing 5-16.

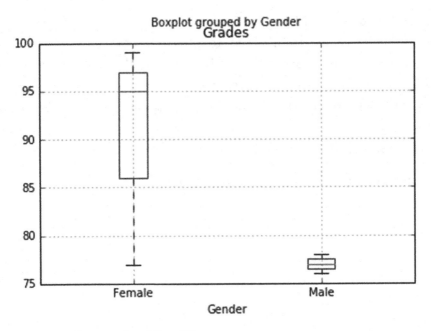

Figure 5-5. *Categorized Box Plot*

Listing 5-16. Categorized Box Plots

And, finally, to adjust the y-axis so that it runs from 0 to 100, we can run the code in Listing 5-17.

Listing 5-17. Adding Code to Adjust the Y-axis

```
axis1 = df.boxplot(by='Gender', column='Grades')
axis1.set_ylim(0,100)
```

It will produce a graph like the one in Figure 5-6.

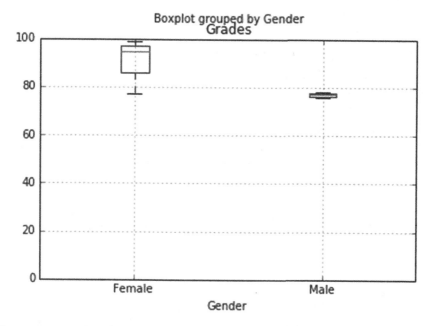

Figure 5-6. *Box Plot Grouped by Gender*

Your Turn

Using the dataset we just created:

- Can you create a box plot of the grades categorized by student status?

- Can you create that box plot with a y-axis that runs from 50 to 110?

Graph a Dataset: Histogram

Because of the nature of histograms, we really need more data than is found in the example dataset we have been working with. Enter the code from Listing 5-18 to import the larger dataset.

Listing 5-18. Importing Dataset from CSV File

```
import matplotlib.pyplot as plt
import pandas as pd
%matplotlib inline
Location = "datasets/gradedata.csv"
df = pd.read_csv(Location)
df.head()
```

To create a simple histogram, we can simply add the code in Listing 5-19.

Listing 5-19. Creating a Histogram not Creating a Box Plot

```
df.hist()
```

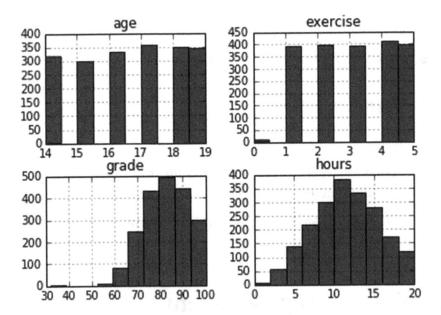

Figure 5-7. *Simple Histogram*

And because pandas is not sure which column you wish to count the values of, it gives you histograms for all the columns with numeric values.

In order to see a histogram for just hours, we can specify it as in Listing 5-20.

Listing 5-20. Creating Histogram for Single Column

```
df.hist(column="hours")
```

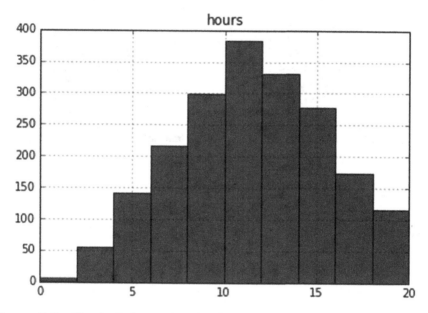

Figure 5-8. *Single Column Histogram*

And to see histograms of hours separated by gender, we can use Listing 5-21.

Listing 5-21. Categorized Histogram

```
df.hist(column="hours", by="gender")
```

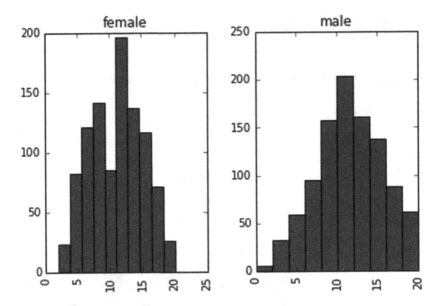

Figure 5-9. *Categorized Histogram*

Your Turn

Can you create an age histogram categorized by gender?

Graph a Dataset: Pie Chart

To create a pie chart, input the code from Listing 5-22.

Listing 5-22. Pie Charting Your Dataset

```
import pandas as pd
import matplotlib.pyplot as plt
%matplotlib inline
names = ['Bob','Jessica','Mary','John','Mel']
absences = [3,0,1,0,8]
detentions = [2,1,0,0,1]
warnings = [2,1,5,1,2]
```

```
GradeList = zip(names,absences,detentions,warnings)
columns=['Names', 'Absences', 'Detentions','Warnings']
df = pd.DataFrame(data = GradeList, columns=columns)
df
```

This code creates a dataset of student rule violations. Next, in a new cell, we will create a column to show the total violations or demerits per student (Listing 5-23).

Listing 5-23. Creating New Column

```
df['TotalDemerits'] = df['Absences'] +
        df['Detentions'] + df['Warnings']
df
```

Finally, to actually create a pie chart of the number of demerits, we can just run the code from Listing 5-24.

Listing 5-24. Creating Pie Chart of Demerits

```
plt.pie(df['TotalDemerits'])
```

Once you run it, you will get a simple pie chart (Figure 5-10).

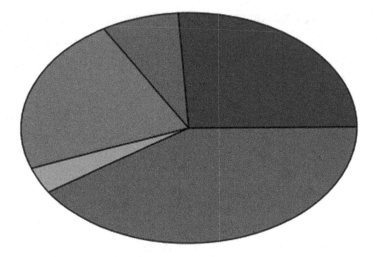

Figure 5-10. *Simple Pie Chart*

But since it is a bit plain (and a bit elongated), let's try the code from Listing 5-25 in a new cell.

Listing 5-25. Creating a Customized Pie Chart

```
plt.pie(df['TotalDemerits'],
        labels=df['Names'],
        explode=(0,0,0,0,0.15),
        startangle=90,
        autopct='%1.1f%%',)
plt.axis('equal')
plt.show()
```

> Line 2: This adds the students' names as labels to the pie pieces.
>
> Line 3: This is what explodes out the pie piece for the fifth student. You can increase or decrease the amount to your liking.

Line 4: This is what rotates the pie chart to different points.

Line 5: This is what formats the numeric labels on the pie pieces.

Line 7: This is what forces the pie to be circular.

And you will see a pie chart that looks like Figure 5-11.

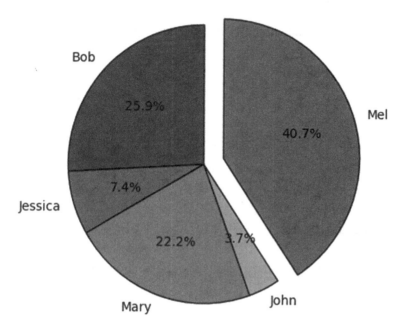

Figure 5-11. *Customized Pie Chart*

Your Turn

What if, instead of highlighting the worst student, we put a spotlight on the best one? Let's rotate the chart and change the settings so we are highlighting John instead of Mel.

Graph a Dataset: Scatter Plot

The code in Listing 5-26 will allow us to generate a simple scatter plot.

Listing 5-26. Creating a Scatter Plot

```
import numpy as np
import pandas as pd
import matplotlib.pyplot as plt
%matplotlib inline
dataframe = pd.DataFrame({'Col':
        np.random.normal(size=200)})
plt.scatter(dataframe.index, dataframe['Col'])
```

> Line 4: specifies that figures should be shown inline
>
> Line 6: generates a random dataset of 200 values
>
> Line 7: creates a scatter plot using the index of the dataframe as the x and the values of column Col as the y

You should get a graph that looks something like Figure 5-12.

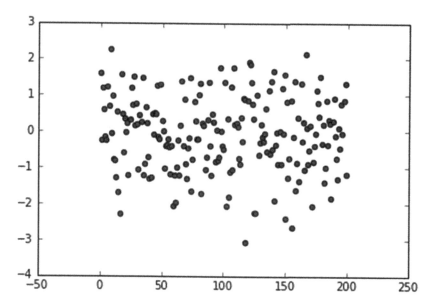

Figure 5-12. *Simple Scatterplot*

Looking at our plot, there doesn't seem to be any pattern to the data. It's random!

Your Turn

Create a scatter plot of the hours and grade data in datasets/gradedata.csv. Do you see a pattern in the data?

## CHAPTER 6

Practice Problems

In this chapter, you will find problems you can use to practice what you have learned. Feel free to use any of the techniques that you have learned, but don't use them all. It would be overkill. Have fun, and good luck!

Analysis Exercise 1

For this exercise, you can find the data in datasets/algebradata.csv.

Frank Mulligrew is the algebra coordinator for Washington, DC public schools. He is required by the school board to gather some statistics. Using the information about his class, calculate the following:

1. Percentage of students with a passing grade

2. Percentage of women with a passing grade

3. Average hours of study for all students

4. Average hours of study for students with a passing grade

© A.J. Henley and Dave Wolf 2018
A.J. Henley and D. Wolf, *Learn Data Analysis with Python*,
https://doi.org/10.1007/978-1-4842-3486-0_6

Analysis Exercise 2

You can find the data in the `datasets/axisdata` file.

Carlos Hugens is the sales manager for Axis Auto Sales, a low-cost regional chain of used car lots. Carlos is getting ready for his annual sales meeting and is looking for the best way to improve his sales group's performance. His data includes the gender, years of experience, sales training, and hours worked per week for each team member. It also includes the average cars sold per month by each salesperson. Find out the following:

1. Average cars sold per month

2. Max cars sold per month

3. Min cars sold per month

4. Average cars sold per month by gender

5. Average hours worked by people selling more than three cars per month

6. Average years of experience

7. Average years of experience for people selling more than three cars per month

8. Average cars sold per month sorted by whether they have had sales training

What do you think is the best indicator of whether someone is a good salesperson?

Analysis Exercise 3

The data can be found in `datasets/dvddata.xlsx`.

Baumgartner DVD's sells high-quality DVD duplicators nationwide. You have just been promoted to sales manager and tasked with analyzing sales trends and making decisions about the best way to handle your sales in the future.

Right now, each of your salespeople covers one or more districts within the same region. Your salespeople contact their customers through either emails, phone calls, or office visits. Emails take about one minute each. Phone calls take about twenty minutes each, and office visits take about two hours. Your staff people work a standard forty-hour week.

1. Figure out the impact of communication methods on number of duplicators sold in a month.

2. Build a model that will predict the number of sales given the number of clients and frequency of each mode of communication.

Analysis Exercise 4

The data can be found in `datasets/tamiami.xlsx`.

Tami, from Miami, wants to open a tamale cart in New York City. She already knows her expenses, but she doesn't know what to charge. She was able to secure the average daily sales data for hot-dog carts by district in the NYC area. Analyze this data to figure out a relationship between price and quantity sold. You can use this relationship as a benchmark for what people are willing to spend for a quick lunch. You need to provide the following:

1. The list of other relevant factors (other than price) that affect sales (if any)

2. The equation for sales quantity

Analysis Project

The data can be found in `datasets/southstyle.xlsx`.

South Carolina–based SouthStyle Foods, a leading manufacturer of sausage, has been selling its products under the brand name SouthStyle for the last 40 years. SouthStyle Foods is engaged in the manufacturing and marketing of high-quality southern-style processed foods such as sausage, bacon, hoppin' john, collard greens, etc. The company provides a perfect blend of traditional southern-style taste tailored to the requirements and preferences of the modern consumer. It combines better taste with natural purity, innovative packaging, and care for health and comes at a reasonable price.

With quality food products and focus on customer satisfaction, SouthStyle Foods maintains a leading position in the processed food section by widening its customer base and making its products available at affordable prices both in South Carolina and nationwide. As a part of its initiatives, the company planned to expand its business to increase the sales of its products in other regions. However, for this, the company wants to know the factors that can increase sales across different states. However, with some new companies coming up, very recently the company witnessed an increase in competition across the industry, resulting in a decrease in its sales.

To discuss the issues, the president, Ashley Sears, called a meeting of the company's senior officers. During a rather lively discussion, they discussed many factors for the fall in sales. However, no common factor emerged. The marketing VP suggested hiring a consultant experienced in business research, and everyone agreed.

SouthStyle Foods hired your marketing research agency, Care Research, for the job. After listening to the problem, your boss thought of using a cross-sectional analysis of the problem, as there are 30 territories from which it must collect data. Your firm started identifying the variables that, according to the company, might have an impact on sales. Based on the collected information (Exhibit I) and the previous studies done, you came up with five important variables that are expected to be crucial in determining the sales. These variables are market potential in the territory, number of shops selling processed foods, number of brokers, number of popular brands in that territory, and population of that territory. The marketing VP wants to know the most important factor or factors to focus on. He also wants to know the likely future demand.

Required Deliverables

1. Identify the most important factors for SouthStyle
    Foods to focus on.

2. Create a formula or model that will allow SouthStyle
    Foods to forecast their sales as they move into new
    territories.

Index

A

Anaconda, 2, 3

B

Bar plot, 74–76
Box plot, 76–79

C

Cleaning data
    calculating and removing
        outliers
        interquartile range (IQR),
          21–22
        standard deviations, 20–21
    description, 19
    filtering inappropriate values,
        24–25
    finding duplicate rows, 26–27
    pandas dataframes, missing
        data, 22–24
    removing punctuation, column
        contents, 27–28
    removing whitespace, column
        contents, 28–29
    SSNs, phone numbers, and zip
        codes, 31–32
    standardizing dates, 29–31
Combining data, multiple excel
    files, 11–13
Computing aggregate statistics
    create dataset, 55–56
    matching rows, 58–59
    measures, central tendency,
        56–57
    sorting data, 59–60
    starting dataset, 57
    variance, 57
Correlation, 60–62

D, E, F, G

Data quality report, 69–71
Datasets
    bar plot, 74–76
    box plot, 76–79
    histograms, 79–82
    line plot, 71–74
    pie chart, 82–86
    scatter plot, 86–87
    unzipping, 4

H

Histograms, 79–82

© A.J. Henley and Dave Wolf 2018
A.J. Henley and D. Wolf, *Learn Data Analysis with Python*,
https://doi.org/10.1007/978-1-4842-3486-0

Get the eBook for only $5!

Why limit yourself?

With most of our titles available in both PDF and ePUB format, you can access your content wherever and however you wish—on your PC, phone, tablet, or reader.

Since you've purchased this print book, we are happy to offer you the eBook for just $5.

To learn more, go to http://www.apress.com/companion or contact support@apress.com.

Apress®

All Apress eBooks are subject to copyright. All rights are reserved by the Publisher, whether the whole or part of the material is concerned, specifically the rights of translation, reprinting, reuse of illustrations, recitation, broadcasting, reproduction on microfilms or in any other physical way, and transmission or information storage and retrieval, electronic adaptation, computer software, or by similar or dissimilar methodology now known or hereafter developed. Exempted from this legal reservation are brief excerpts in connection with reviews or scholarly analysis or material supplied specifically for the purpose of being entered and executed on a computer system, for exclusive use by the purchaser of the work. Duplication of this publication or parts thereof is permitted only under the provisions of the Copyright Law of the Publisher's location, in its current version, and permission for use must always be obtained from Springer. Permissions for use may be obtained through RightsLink at the Copyright Clearance Center. Violations are liable to prosecution under the respective Copyright Law.

Printed in the United States
By Bookmasters